The Sharing Ideas from What If We Get It Right Workbook

Learning Points from Ayana Elizabeth Johnson's Book

Novo Leno

Disclaimer

The content in this workbook is provided for educational and informational purposes only. It is intended to support and enhance your understanding of the concepts presented in the book "What If We Get It Right?" by Ayana Elizabeth Johnson. The author, Novo Leno, and the publisher make no guarantees regarding the accuracy, completeness, or applicability of the information contained herein.

This workbook is not intended as a substitute for professional advice or guidance. Readers are encouraged to consult with appropriate professionals for advice tailored to their individual circumstances. The author and publisher disclaim any liability for any loss or damage incurred by the use or reliance on the information provided in this workbook.

Table of content

Chapter 1: Introduction to Visionary Thinking

1.1 What is Visionary Thinking?

Visionary thinking is an advanced cognitive skill that allows individuals to conceptualize the future and develop innovative solutions that go beyond the status quo. Unlike reactive thinking, which focuses on addressing immediate problems, visionary thinking encourages individuals to look ahead and imagine what might be possible. This skill is critical in various fields, from technology and medicine to education and, importantly, climate change.

Visionary thinkers use imagination as a tool to forecast future challenges, opportunities, and solutions. They are not confined by existing paradigms; instead, they often challenge and redefine them. Visionary thinking involves creativity, strategic planning, and a deep understanding of the larger systems at play, whether they be social, economic, or environmental.

In the context of climate change, visionary thinking is essential because it encourages us to envision new ways of living, consuming, and interacting with the planet. While traditional methods of combating climate change have focused on mitigation and adaptation, visionary thinking allows us to explore transformative solutions that address the root causes of environmental degradation and pave the way for a sustainable future.

Visionary thinking has been a driving force behind many of the world's most significant advancements. Historical figures such as Leonardo da Vinci, who envisioned flying machines centuries before airplanes were invented, or Nikola Tesla, whose concepts of wireless communication predated modern technologies, demonstrate how visionary thinking can lead to groundbreaking innovations.

In the climate context, individuals like Greta Thunberg, who has sparked a global movement for climate action, and Elon Musk, whose work with Tesla and SpaceX reimagines transportation and energy systems, embody visionary thinking. These figures illustrate how the power of imagination, when combined with practical action, can reshape the world.

Characteristics of Visionary Thinking:

1. **Long-Term Focus**: Visionary thinkers prioritize long-term outcomes over immediate results. They are not satisfied with

temporary fixes but strive to create lasting change.

2. **Creativity**: Imagination is at the core of visionary thinking. It involves envisioning possibilities that do not yet exist and thinking beyond current limitations.

3. **Optimism**: Visionary thinkers tend to be optimistic about the future, even when faced with significant challenges. This optimism fuels their determination to explore innovative solutions.

4. **Risk-Taking**: Because visionary thinking involves proposing new and often unconventional ideas, it requires a certain level of risk-taking. Visionaries are not afraid to fail because they understand that failure is often a necessary step toward success.

5. **Systems Thinking**: Visionary thinkers have a holistic understanding of the world. They recognize that individual actions are part of larger systems, and they think strategically about how to influence these systems for positive change.

1.2 The Role of Imagination in Problem-Solving

Imagination plays a central role in visionary thinking. It allows us to transcend the boundaries of what is currently possible and explore new ideas, perspectives, and solutions. Imagination is not just about fantasizing; it is a practical tool for problem-solving that enables us to foresee potential futures and develop innovative strategies to address complex challenges.

In problem-solving, imagination helps us break free from conventional approaches. When faced with a difficult problem, it can be easy to fall into the trap of trying to solve it using the same methods that have been used in the past. However, visionary thinkers understand that true innovation comes from stepping outside of these boundaries and exploring new possibilities.

Take, for instance, the challenge of transitioning to renewable energy sources. Traditional approaches may focus on expanding existing infrastructure or improving efficiency. However, imaginative thinkers have proposed solutions such as **floating solar farms** or **space-based solar power**, ideas that seemed fantastical just a few decades ago. Yet, as technology advances and the urgency of climate action grows, these once imaginative ideas are becoming viable options.

Imagination also allows us to foresee the potential consequences of our actions. In the context of climate change, it is essential to imagine not only the technical solutions but also the social, economic, and political implications of those solutions. For example, transitioning to a low-carbon economy will have far-reaching impacts on industries, employment, and global trade. Visionary thinkers must anticipate these impacts and propose ways to mitigate any negative consequences while maximizing the positive outcomes.

How Imagination Enhances Problem-Solving:

1. **Expands Possibilities**: Imagination allows us to think beyond conventional solutions and explore a broader range of possibilities. This is essential in addressing complex, multi-faceted problems like climate change.

2. **Foresee Consequences**: Imagination enables us to anticipate the potential consequences of our actions, both positive and negative, allowing us to make more informed decisions.

3. **Facilitates Collaboration**: Imaginative solutions often require collaboration between diverse groups of people, each bringing their own perspectives and expertise to the table. Visionary thinkers are adept at bringing these groups together to work toward a common goal.

4. **Inspires Action**: Imaginative ideas can be incredibly motivating, inspiring people to take action and work toward a shared vision of the future.

1.3 Key Characteristics of Visionary Thinkers

Visionary thinkers possess several key traits that enable them to see beyond the present and envision a better future. These traits are not innate; they can be cultivated through practice, reflection, and a commitment to learning.

1.3.1 Curiosity

Curiosity is the driving force behind visionary thinking. Visionary thinkers are constantly asking questions, seeking new information, and exploring different perspectives. They are not content with the status quo and are always looking for ways to improve and innovate.

Curiosity fuels the desire to learn and grow, which is essential in a rapidly changing world. In the context of climate change, curiosity leads to a deeper understanding of the complex systems at play, from the natural environment to human behavior. Visionary thinkers ask questions like: How can we redesign cities to be more sustainable? What new technologies can help us reduce greenhouse gas emissions? How can we shift cultural attitudes toward environmental responsibility?

1.3.2 Creativity

Creativity is at the heart of visionary thinking. It involves the ability to generate new ideas, connect seemingly unrelated concepts, and think outside the box. Visionary thinkers are not bound by conventional wisdom; they are willing to explore unconventional solutions to complex problems.

In climate solutions, creativity has led to the development of cutting-edge technologies like **carbon capture and storage** (CCS) and **regenerative agriculture** practices that restore ecosystems while producing food. Visionary thinkers are constantly pushing the boundaries of what is possible, reimagining everything from transportation systems to energy grids.

1.3.3 Persistence

Visionary thinking requires persistence. The path to innovation is often filled with obstacles, setbacks, and failures. However, visionary thinkers are not discouraged by these challenges; instead, they view them as opportunities to learn and grow.

Persistence is particularly important in the fight against climate change, where progress can sometimes feel slow and incremental. Visionary thinkers understand that meaningful change takes time and that setbacks are a natural part of the process. They remain committed to their vision, even in the face of adversity, because they believe in the long-term benefits of their work.

1.3.4 Empathy

Empathy is a critical trait for visionary thinkers, particularly in the context of climate action. Visionary thinkers are not only concerned with the technical aspects of a problem; they also consider the human impact of their solutions. They strive to create inclusive, equitable, and just solutions that benefit all members of society.

In climate action, empathy leads to a deeper understanding of the social and economic disparities that contribute to environmental degradation. Visionary thinkers recognize that vulnerable communities are often the most affected by climate change and are committed to creating solutions that address these inequities.

Empathy also allows visionary thinkers to build strong relationships and collaborate effectively with others. By understanding the needs and perspectives of different stakeholders, visionary thinkers can create solutions that are more likely to be embraced and implemented.

1.4 Visionary Thinking in Climate Solutions

The climate crisis presents one of the most significant challenges humanity has ever faced. The consequences of inaction are dire, but the potential for positive change is immense. Visionary thinking is essential for unlocking this potential and creating a sustainable future for all.

While traditional approaches to climate action focus on mitigating the impacts of climate change, visionary thinking goes further by reimagining the systems that have led to environmental degradation in the first place. This involves not only reducing greenhouse gas emissions but also transforming the way we live, work, and interact with the planet.

1.4.1 Innovative Climate Solutions

Visionary thinkers are already at the forefront of developing innovative solutions to address climate change. Some of these solutions are technological, while others involve rethinking social and economic systems.

- **Renewable Energy**: Visionary thinkers are leading the transition to renewable energy sources like solar, wind, and hydropower. They are also exploring new technologies, such as **fusion energy** and **space-based solar power**, that have the potential to revolutionize the way we produce and consume energy.
- **Circular Economy**: The concept of a circular economy, where resources are reused and recycled rather than discarded, is gaining traction among visionary thinkers. This approach reduces waste and conserves natural resources while creating economic opportunities.

- **Sustainable Agriculture**: Visionary thinkers are developing new agricultural practices that promote sustainability and resilience. **Regenerative agriculture**, for example, focuses on restoring soil health and biodiversity while producing food. This approach not only sequesters carbon but also improves the resilience of ecosystems to climate change.

1.4.2 Reimagining Urban Environments

Cities are responsible for a significant portion of global greenhouse gas emissions, making them a key focus for visionary climate solutions. Reimagining urban environments involves creating cities that are not only more sustainable but also more livable and resilient.

- **Green Infrastructure**: Visionary urban planners are incorporating green infrastructure into city designs. This includes features like **green roofs**, **urban forests**, and **rain gardens** that help manage stormwater, reduce urban heat islands, and improve air quality.

- **Energy-Efficient Buildings**: The design and construction of energy-efficient buildings are essential for reducing the carbon footprint of cities. Visionary architects are developing **passive house** designs that minimize energy use through careful design and construction techniques.

1.4.3 Transformative Social Change

Addressing climate change also requires transformative social change. Visionary thinkers are exploring ways to shift cultural attitudes and behaviors toward greater environmental responsibility.

- **Education and Awareness:** Increasing awareness about climate change and its impacts is crucial for fostering a culture of environmental stewardship. Visionary educators and activists are developing innovative educational programs and campaigns to engage people in climate action.
- **Policy and Advocacy:** Visionary thinkers are working to influence policy and advocate for systemic change. This includes promoting policies that support renewable energy, conservation, and climate justice.

1.5 Exercises and Reflections

To cultivate visionary thinking and apply it to climate solutions, it is essential to engage in activities and reflections that challenge conventional thinking and inspire creativity.

1.5.1 Activities to Develop Visionary Thinking

- **Brainstorming Sessions**: Organize brainstorming sessions with a diverse group of people to explore new ideas and solutions. Encourage participants to think outside the box and consider unconventional approaches.
- **Scenario Planning**: Develop and analyze different future scenarios related to climate change. Consider the potential impacts of various trends and actions, and use these scenarios to guide decision-making.
- **Vision Boards**: Create vision boards that represent your goals and aspirations for a sustainable future. Use images, words, and

symbols to depict your vision and keep it visible as a source of inspiration.

1.5.2 Reflection Questions

- What are some areas in your life or work where you could apply visionary thinking?
- How can you challenge existing assumptions and explore new possibilities?
- Who are some visionary thinkers you admire, and what can you learn from their approaches?
- How can you incorporate visionary thinking into your daily actions and decisions?

Chapter 2: Exploring Climate Solutions

2.1 Overview of Climate Solutions

Climate solutions encompass a wide range of strategies and technologies aimed at mitigating the impacts of climate change and adapting to its effects. These solutions can be categorized into several key areas, including technological innovations, policy measures, and community-based initiatives. Each category plays a crucial role in addressing the climate crisis and contributes to building a sustainable future.

The urgency of climate action is underscored by the increasing frequency and severity of extreme weather events, rising sea levels, and disruptions to ecosystems and communities. Effective climate solutions must therefore address both the root causes of climate change, such as greenhouse gas emissions, and the adaptation needs of vulnerable communities.

This chapter explores some of the most promising climate solutions in detail, providing an overview of their potential impacts, benefits, and challenges.

2.2 Technological Innovations

Technological innovations are at the forefront of climate solutions, offering new ways to reduce emissions, improve energy efficiency, and enhance environmental sustainability. These innovations span various sectors, including energy, transportation, and agriculture.

2.2.1 Renewable Energy Technologies

Renewable energy technologies are essential for reducing reliance on fossil fuels and decreasing greenhouse gas emissions. Key renewable energy sources include solar power, wind power, hydroelectric power, and geothermal energy.

- **Solar Power**: Solar energy harnesses the power of the sun to generate electricity. Advancements in photovoltaic (PV) technology have led to more efficient and affordable solar panels. Innovations such as **solar roof tiles** and **floating solar farms** are expanding the potential applications of solar power. Solar power has the advantage of being scalable and adaptable to various settings, from residential rooftops to large-scale solar farms.
- **Wind Power**: Wind turbines convert the kinetic energy of wind

into electricity. Modern wind turbine designs have become more efficient, with advancements such as **offshore wind farms** and **vertical-axis wind turbines**. Offshore wind farms, located in bodies of water, benefit from stronger and more consistent winds, making them a promising solution for large-scale renewable energy production.

- **Hydroelectric Power**: Hydroelectric power generates electricity from the movement of water, typically through dams. While traditional hydroelectric dams can have environmental impacts, innovative approaches like **pumped-storage hydropower** and **run-of-river systems** offer more sustainable alternatives. Pumped-storage hydropower helps balance supply and demand by storing energy during periods of low demand and releasing it during peak periods.

- **Geothermal Energy**: Geothermal energy utilizes heat from the Earth's interior to generate electricity and provide heating. Advances in **geothermal drilling** and **enhanced geothermal systems (EGS)** are expanding the potential of geothermal energy. Geothermal energy is reliable and provides a continuous source of power, with minimal environmental impact compared to fossil fuels.

2.2.2 Energy Storage Solutions

Energy storage technologies are crucial for managing the intermittent nature of renewable energy sources. Effective storage solutions can enhance the reliability and stability of the energy grid.

- **Batteries**: Lithium-ion batteries are widely used for energy storage due to their high energy density and efficiency. However, research is ongoing into alternative battery technologies, such as **solid-state batteries** and **flow batteries**, which offer potential advantages in terms of safety, cost, and performance.
- **Pumped Hydro Storage**: Pumped hydro storage involves using excess electricity to pump water to a higher elevation. The stored water is then released to generate electricity when needed. This technology is well-established and provides large-scale energy

storage, but it requires specific geographic conditions.

- **Compressed Air Energy Storage (CAES)**: CAES stores energy by compressing air and storing it in underground caverns or tanks. The compressed air is released to drive turbines and generate electricity. CAES systems can provide long-duration energy storage and help stabilize the grid.

2.2.3 Sustainable Agriculture Practices

Agriculture is both a source of greenhouse gas emissions and a potential solution for climate change. Sustainable agriculture practices aim to reduce emissions, improve soil health, and enhance food security.

- **Regenerative Agriculture**: Regenerative agriculture focuses on restoring soil health and biodiversity through practices such as **cover cropping, no-till farming**, and **rotational grazing**. These practices improve soil carbon sequestration and increase resilience to extreme weather events.

- **Agroforestry**: Agroforestry integrates trees and shrubs into agricultural landscapes, providing multiple benefits such as carbon sequestration, soil erosion control, and enhanced biodiversity. Agroforestry systems can improve farm productivity and sustainability while contributing to climate goals.

- **Precision Agriculture**: Precision agriculture uses technology such as **drones, GPS**, and **sensors** to optimize the use of resources and reduce environmental impacts. By targeting inputs such as water and fertilizers more precisely, precision agriculture can reduce waste and improve crop yields.

2.3 Policy Measures

Effective climate policies are essential for driving the transition to a low-carbon economy and promoting sustainable practices. Policy measures can be implemented at local, national, and international levels to address various aspects of climate change.

2.3.1 Carbon Pricing

Carbon pricing is a market-based approach to reducing greenhouse gas emissions by assigning a cost to carbon emissions. There are two main types of carbon pricing mechanisms:

- **Carbon Tax**: A carbon tax places a direct price on carbon emissions, providing an economic incentive for businesses and individuals to reduce their carbon footprint. The revenue generated from carbon taxes can be used to fund renewable energy projects, support vulnerable communities, and invest in climate adaptation measures.

- **Cap-and-Trade**: Cap-and-trade systems set a cap on total emissions and allow businesses to buy and sell emissions permits. This market-based approach creates financial incentives for companies to reduce their emissions and invest in cleaner technologies. The cap is gradually reduced over time, leading to overall emissions reductions.

2.3.2 Renewable Energy Incentives

Governments can promote the adoption of renewable energy technologies through various incentives and subsidies. These measures can include:

- **Feed-in Tariffs**: Feed-in tariffs guarantee a fixed payment for renewable energy producers, providing financial stability and encouraging investment in clean energy projects.
- **Tax Credits and Rebates**: Tax credits and rebates can reduce the upfront costs of renewable energy installations, making them more accessible to individuals and businesses. Examples include the **Investment Tax Credit (ITC)** and **Production Tax Credit (PTC)** in the United States.
- **Renewable Energy Standards**: Renewable energy standards

mandate a certain percentage of electricity to come from renewable sources. These standards can drive the development of renewable energy projects and increase the share of clean energy in the electricity mix.

2.3.3 Climate Adaptation Policies

Climate adaptation policies focus on preparing for and responding to the impacts of climate change. These policies can include:

- **Infrastructure Resilience**: Investing in resilient infrastructure, such as **flood defenses**, **heat-resistant buildings**, and **drought-tolerant transportation systems**, can help communities adapt to climate impacts and reduce vulnerability.

- **Disaster Preparedness**: Developing disaster preparedness plans and early warning systems can enhance community resilience and reduce the impacts of extreme weather events.
- **Ecosystem Protection**: Protecting and restoring ecosystems, such as **wetlands**, **mangroves**, and **coral reefs**, can enhance natural resilience to climate impacts and provide essential services such as flood protection and carbon sequestration.

2.4 Community-Based Initiatives

Community-based initiatives are crucial for implementing climate solutions at the local level and engaging individuals in climate action. These initiatives often involve grassroots organizations, local governments, and community members working together to address climate challenges and promote sustainability.

2.4.1 Grassroots Movements

Grassroots movements play a vital role in raising awareness about climate change and advocating for local and global action. These movements can include:

- **Climate Strikes**: Organized by groups like **Fridays for Future**, climate strikes mobilize millions of people around the world to demand urgent action on climate change. These events highlight the importance of climate action and put pressure on policymakers to address environmental issues.
- **Community Gardens**: Community gardens promote local food production and provide educational opportunities for

residents to learn about sustainable agriculture. These gardens can enhance food security, reduce transportation emissions, and foster community engagement.

- **Local Climate Action Plans**: Many communities are developing local climate action plans that outline specific goals and strategies for reducing emissions and improving resilience. These plans often involve collaboration between local governments, businesses, and residents.

2.4.2 Sustainable Business Practices

Businesses play a significant role in driving sustainability and reducing their environmental impact. Sustainable business practices can include:

- **Corporate Social Responsibility (CSR):** CSR initiatives focus on the social and environmental impacts of business operations. Companies that prioritize CSR often implement practices such as **sustainable sourcing, energy efficiency,** and **waste reduction.**
- **Green Certifications:** Green certifications, such as **LEED** (Leadership in Energy and Environmental Design) and **B Corp**, recognize businesses that meet high standards of environmental and social performance. These certifications can help businesses demonstrate their commitment to sustainability and attract environmentally conscious consumers.
- **Innovation and Collaboration:** Many businesses are investing in

research and development to create innovative solutions for climate challenges. Collaboration with other organizations, including NGOs and government agencies, can enhance the effectiveness of these efforts and drive collective action.

2.5 Challenges and Opportunities

While climate solutions offer immense potential for addressing environmental challenges, they also face several obstacles. Understanding these challenges and identifying opportunities for overcoming them is essential for advancing climate action.

2.5.1 Challenges

- **Funding and Investment**: Many climate solutions require significant investment, and securing funding can be a barrier to implementation. Governments, private sector investors, and philanthropic organizations must work together to provide the necessary financial support.

- **Technological Limitations**: Some climate technologies are still in the development stage and may face technical or scalability challenges. Continued research and innovation are needed to advance these technologies and make them widely available.
- **Political and Social Resistance**: Climate action often faces resistance from political or social groups that may be affected by changes in policies or practices. Building broad-based support and addressing concerns through

dialogue and education are crucial for overcoming resistance.

2.5.2 Opportunities

- **Job Creation**: The transition to a low-carbon economy presents opportunities for job creation in sectors such as renewable energy, energy efficiency, and sustainable agriculture. Investing in these sectors can drive economic growth while addressing climate challenges.
- **Technological Advancements**: Rapid advancements in technology offer new solutions for reducing emissions and enhancing sustainability. Continued innovation and adoption of emerging technologies can drive significant progress in climate action.

- **Global Collaboration**: Addressing climate change requires global cooperation and collaboration. International agreements, such as the **Paris Agreement**, provide a framework for countries to work together and share knowledge and resources.

2.6 Exercises and Reflections

Engaging in exercises and reflections can help deepen your understanding of climate solutions and inspire action. Consider the following activities to explore climate solutions further:

2.6.1 Activities to Explore Climate Solutions

- **Research and Analysis**: Conduct research on specific climate solutions and analyze their potential impacts, benefits, and challenges. Consider how these solutions could be applied in your community or industry.

- **Case Studies**: Review case studies of successful climate initiatives and projects. Identify key factors that contributed to their success and explore how similar approaches could be implemented in other contexts.
- **Solution Mapping**: Create a visual map of climate solutions, categorizing them by type (technological, policy, community-based) and potential impact. Use this map to identify gaps and opportunities for further action.

2.6.2 Reflection Questions

- What are some of the most promising climate solutions you

have encountered, and why do you find them compelling?

- How can you contribute to advancing climate solutions in your personal or professional life?
- What challenges might you face in implementing climate solutions, and how can you address them?
- Who are the key stakeholders involved in climate solutions, and how can you engage with them to support their efforts?

Chapter 3: Understanding the Role of Policy and Culture

3.1 The Intersection of Policy and Climate Action

Policy plays a pivotal role in shaping climate action and environmental sustainability. Effective policies can drive significant progress toward climate goals by creating frameworks for reducing emissions, promoting renewable energy, and fostering innovation. This section examines how different policy approaches impact climate action and the factors that contribute to successful policy implementation.

3.1.1 Policy Frameworks for Climate Action

Several key policy frameworks are essential for addressing climate change and promoting sustainability. These frameworks provide guidelines for governments, businesses, and individuals to follow in their efforts to mitigate and adapt to climate impacts.

- **Climate Agreements and Treaties**: International climate agreements, such as the **Paris Agreement**, set global targets for reducing greenhouse gas emissions and fostering climate resilience. These agreements require countries to develop and implement national climate action plans, known as **Nationally Determined Contributions (NDCs)**, to meet their emission reduction commitments.
- **Carbon Pricing Mechanisms**: Carbon pricing mechanisms, including **carbon taxes** and **cap-and-trade systems**, provide economic incentives for reducing

emissions. By assigning a price to carbon emissions, these mechanisms encourage businesses and individuals to adopt cleaner technologies and practices.

- **Renewable Energy Standards**: Renewable energy standards mandate a certain percentage of energy to come from renewable sources. These standards can drive the development of renewable energy projects and increase the share of clean energy in the electricity mix. For example, **Renewable Portfolio Standards (RPS)** require utilities to obtain a specified percentage of their energy from renewable sources.
- **Energy Efficiency Policies**: Energy efficiency policies aim to reduce energy consumption and improve the performance of energy

systems. Policies such as **Building Codes and Standards** and **Energy Efficiency Resource Standards (EERS)** set requirements for energy-efficient construction, appliances, and industrial processes.

3.1.2 Policy Implementation and Challenges

Implementing climate policies effectively requires addressing various challenges and overcoming obstacles. Successful policy implementation depends on several factors, including political will, stakeholder engagement, and adequate resources.

- **Political Will and Leadership**: Effective climate policies require strong political will and leadership. Policymakers must be committed to prioritizing climate action and addressing opposition from vested interests. Leadership at various levels—local, national, and international—can drive policy change and mobilize resources.

- **Stakeholder Engagement**: Engaging stakeholders, including businesses, community organizations, and the public, is crucial for successful policy implementation. Stakeholder input can help shape policies that are practical and effective. Public support and involvement can also enhance the legitimacy and success of climate policies.
- **Resources and Funding**: Adequate resources and funding are essential for implementing and enforcing climate policies. Governments and organizations must allocate financial resources to support policy measures,

conduct research, and develop infrastructure. Funding mechanisms, such as **green bonds** and **climate finance**, can help support policy implementation.

- **Monitoring and Evaluation:** Monitoring and evaluating the effectiveness of climate policies are important for ensuring that they achieve their intended goals. Regular assessment can identify areas for improvement and inform future policy adjustments. Monitoring systems can track progress toward emission reduction targets and evaluate the impact of policies on environmental and social outcomes.

3.2 Cultural Influences on Climate Action

Culture plays a significant role in shaping attitudes and behaviors related to climate action. Cultural values, norms, and practices influence how individuals and communities perceive and respond to environmental challenges. Understanding cultural influences can help design effective climate strategies and foster a culture of sustainability.

3.2.1 Cultural Values and Environmental Attitudes

Cultural values and beliefs shape how people view and interact with the environment. Different cultures have varying perspectives on nature, resource use, and environmental stewardship.

- **Environmental Ethics**: Environmental ethics refers to the moral principles guiding human interactions with the natural world. Cultural traditions and religious beliefs often shape these ethical views. For example, some cultures emphasize the intrinsic value of nature and advocate for protecting ecosystems as a moral obligation.

- **Resource Use and Conservation**: Cultural attitudes toward resource use and conservation influence behaviors such as consumption, waste management, and conservation practices. Cultures with strong traditions of resource conservation may prioritize sustainable practices and conservation efforts.
- **Community and Social Norms**: Social norms and community practices can impact environmental behavior and climate action. Communities with a strong emphasis on collective

well-being and environmental stewardship may have higher levels of engagement in climate initiatives.

3.2.2 Promoting a Culture of Sustainability

Creating a culture of sustainability involves fostering attitudes and behaviors that support environmental protection and climate action. Several strategies can help promote sustainability within communities and organizations.

- **Education and Awareness:** Education plays a crucial role in raising awareness about climate change and its impacts. Educational programs and campaigns can help individuals understand the importance of sustainability and motivate them to take action. Schools, universities, and community organizations can all contribute to environmental education.

- **Cultural Celebrations and Traditions**: Incorporating sustainability into cultural celebrations and traditions can reinforce environmental values and practices. For example, communities can adopt sustainable practices during festivals and events, such as using eco-friendly materials and reducing waste.
- **Role Models and Leadership**: Role models and leaders who demonstrate a commitment to sustainability can inspire others to follow suit. Public figures, community leaders, and organizations that prioritize environmental stewardship can

serve as powerful examples and advocates for climate action.

- **Collaboration and Partnerships**: Collaborating with diverse stakeholders, including businesses, NGOs, and government agencies, can enhance the effectiveness of sustainability initiatives. Partnerships can leverage resources, expertise, and networks to address climate challenges and promote sustainable practices.

3.3 Case Studies of Policy and Cultural Impact

Examining case studies of successful climate policies and cultural initiatives can provide valuable insights into effective strategies and practices. These case studies highlight the role of policy and culture in driving climate action and achieving sustainability goals.

3.3.1 Successful Climate Policies

- **Sweden's Carbon Tax**: Sweden implemented a carbon tax in 1991, becoming one of the first countries to adopt such a policy. The tax has successfully reduced greenhouse gas emissions while supporting economic growth. Sweden's experience demonstrates the effectiveness of carbon pricing in driving emissions reductions and promoting renewable energy.

- **California's Cap-and-Trade System**: California's cap-and-trade system, established in 2013, sets a cap on statewide emissions and allows businesses to trade emissions permits. The system has helped reduce emissions while generating revenue for climate initiatives. California's approach highlights the potential of market-based mechanisms for achieving climate goals.

- **Germany's Energiewende**: Germany's Energiewende (energy transition) policy aims to shift the country toward renewable energy and reduce greenhouse gas emissions. The policy includes measures such as feed-in tariffs, renewable energy standards, and energy efficiency initiatives. Germany's experience showcases the importance of comprehensive policies in driving the transition to a low-carbon economy.

3.3.2 Cultural Initiatives and Engagement

- **The Zero Waste Movement**: The Zero Waste movement promotes the goal of reducing waste to the minimum through practices such as recycling, composting, and reducing consumption. Community-based initiatives and cultural campaigns have helped raise awareness and encourage individuals and organizations to adopt zero waste practices.

- **Eco-Friendly Festivals**: Various festivals around the world have adopted eco-friendly practices to minimize their environmental impact. Examples include **Glastonbury Festival** in the UK, which implements waste reduction and recycling programs, and the **Burning Man Festival** in the US, which promotes sustainability and leave-no-trace principles.
- **Traditional Ecological Knowledge (TEK)**: Indigenous cultures have long practiced sustainable environmental management through traditional

ecological knowledge. TEK includes practices such as **sustainable hunting** and **agroecology** that contribute to conservation and resource management. Recognizing and integrating TEK can enhance contemporary climate strategies.

3.4 Exercises and Reflections

Engaging in exercises and reflections can help deepen your understanding of the role of policy and culture in climate action. Consider the following activities to explore these topics further:

3.4.1 Activities to Explore Policy and Culture

- **Policy Analysis**: Analyze existing climate policies at local, national, or international levels. Evaluate their effectiveness, identify strengths and weaknesses, and consider potential improvements or alternative approaches.

- **Cultural Research**: Research cultural attitudes and practices related to environmental sustainability in different regions or communities. Explore how cultural values influence environmental behaviors and identify opportunities for promoting sustainability within specific cultural contexts.
- **Stakeholder Interviews**: Conduct interviews with stakeholders involved in climate policy or sustainability initiatives. Gather insights on their experiences, challenges, and strategies for advancing climate action. Use this information to inform your understanding of policy and cultural impacts.

3.4.2 Reflection Questions

- How do different policy approaches impact climate action and sustainability? What factors contribute to the success or failure of climate policies?
- In what ways do cultural values and norms influence environmental attitudes and behaviors? How can cultural factors be leveraged to promote sustainability?
- What are some examples of successful climate policies or cultural initiatives that have inspired you? What lessons can you apply to your own climate efforts?
- How can you engage with policymakers, community leaders, and other stakeholders to support

climate action and sustainability
within your own context?

Chapter 4: Insights from Visionary Voices

4.1 The Power of Visionary Voices

Visionary voices play a crucial role in shaping our understanding of climate change and inspiring action. These individuals bring fresh perspectives, innovative ideas, and a deep commitment to addressing environmental challenges. Their insights offer valuable guidance for navigating the complex landscape of climate action and sustainability. This chapter delves into the contributions of some of these visionary voices, highlighting their perspectives on climate solutions, social justice, and the future of our planet.

4.1.1 Influential Thinkers and Activists

Several influential thinkers and activists have made significant contributions to climate discourse. Their work spans various fields, including science, policy, art, and activism. By examining their contributions, we gain a broader understanding of the diverse approaches to climate action.

- **Paola Antonelli**: As a curator and designer, Paola Antonelli explores the intersection of design and climate action. Her work emphasizes the importance of creativity and innovation in addressing environmental challenges. Antonelli's exhibitions and projects highlight how design can drive sustainable solutions and inspire change.
- **Xiye Bastida**: A climate activist and member of the Otomi-Toltec community, Xiye Bastida advocates for indigenous rights

and climate justice. Her activism focuses on amplifying the voices of marginalized communities and promoting equitable climate solutions. Bastida's work underscores the need for inclusive approaches to climate action that consider the perspectives of those most affected by environmental issues.

- **Wendell Berry**: An acclaimed author and farmer, Wendell Berry advocates for sustainable agriculture and local resilience. His writings emphasize the importance of nurturing a connection to the land and promoting regenerative farming practices. Berry's perspective highlights the role of traditional knowledge and community-based approaches in fostering environmental stewardship.

- **Kate Orff**: An urban designer and landscape architect, Kate Orff's work explores how design can address climate adaptation and resilience. Her projects focus on integrating natural systems into urban environments to enhance sustainability and mitigate climate impacts. Orff's approach emphasizes the potential of design to create more livable and resilient cities.

4.1.2 Insights from Diverse Perspectives

The perspectives of visionary voices offer diverse insights into climate solutions and sustainability. By exploring these insights, we can gain a deeper understanding of the multifaceted nature of climate action and the ways in which different approaches can contribute to a more sustainable future.

- **Erica Deeman**: A photographer and artist, Erica Deeman's work addresses themes of identity and environmental change. Her art challenges viewers to consider the human impact of climate change and the importance of diversity in environmental advocacy. Deeman's perspective highlights the role of art in raising awareness and fostering dialogue about climate issues.
- **Jigar Shah**: A clean energy entrepreneur and advocate, Jigar Shah focuses on promoting renewable energy technologies and innovative financing models. His work emphasizes the potential

of market-based solutions and entrepreneurial approaches to drive the transition to a low-carbon economy. Shah's insights highlight the importance of scaling up clean energy solutions and overcoming financial barriers.

- **Régine Clément**: An environmental advocate and researcher, Régine Clément's work explores the intersection of climate change and social justice. Her research highlights the disproportionate impacts of climate change on marginalized communities and the need for equitable solutions. Clément's perspective underscores the importance of addressing social inequalities in climate action efforts.
- **Mustafa Suleyman**: As a co-founder of DeepMind, Mustafa

Suleyman explores the potential of artificial intelligence to address climate challenges. His work focuses on leveraging AI for climate modeling, energy optimization, and resource management. Suleyman's insights emphasize the role of technology in advancing climate science and improving decision-making.

4.2 Contributions to Climate Solutions

The contributions of visionary voices extend beyond their individual perspectives. Their work often involves collaborative efforts, innovative projects, and advocacy that drive meaningful progress in climate action. This section highlights some of the key contributions of these visionary voices and their impact on climate solutions.

4.2.1 Collaborative Initiatives

Many visionary voices engage in collaborative initiatives that bring together diverse stakeholders to address climate challenges. These initiatives often involve partnerships between scientists, policymakers, activists, and community leaders.

- **The Sunrise Movement**: Founded by a group of young activists, the Sunrise Movement advocates for bold climate action and a Green New Deal. The movement's efforts include grassroots organizing, policy advocacy, and public education. Visionary voices within the movement work to mobilize communities and influence policy at the national level.
- **The Drawdown Project**: The Drawdown Project, led by Paul Hawken, focuses on identifying

and promoting climate solutions that can achieve significant reductions in greenhouse gas emissions. The project's research highlights practical strategies for addressing climate change, including renewable energy, carbon sequestration, and sustainable agriculture.

- **The Climate Reality Project**: Founded by former Vice President Al Gore, the Climate Reality Project works to raise awareness about climate change and advocate for policy solutions. The project's initiatives include training climate leaders, organizing campaigns, and promoting climate education. Visionary voices involved in the project contribute to its efforts to inspire global climate action.

4.2.2 Innovative Projects and Solutions

Visionary voices often spearhead innovative projects and solutions that address climate challenges in unique ways. These projects demonstrate the potential for creative approaches to drive progress and create impact.

- **The Ocean Cleanup**: Founded by Boyan Slat, The Ocean Cleanup project aims to remove plastic pollution from the world's oceans. The project uses advanced technologies, such as floating barriers and autonomous systems, to collect and remove plastic waste. Slat's innovative approach highlights the potential for technology to address environmental pollution.
- **The B Team**: The B Team, co-founded by Richard Branson and Jochen Zeitz, is a nonprofit

organization focused on advancing sustainable business practices. The organization's initiatives include promoting responsible corporate behavior, advocating for climate action, and supporting sustainable development goals. The B Team's work emphasizes the role of business leadership in driving positive change.

- **Regenerative Agriculture Initiatives**: Various regenerative agriculture initiatives focus on restoring soil health, increasing biodiversity, and sequestering carbon through sustainable farming practices. These initiatives, led by organizations such as the Regenerative Agriculture Initiative and the Soil Carbon Coalition, demonstrate the potential for agriculture to contribute to climate solutions.

4.3 Practical Insights and Recommendations

The insights from visionary voices offer practical recommendations for individuals, organizations, and policymakers seeking to advance climate action. This section provides actionable advice based on the contributions and perspectives of these thought leaders.

4.3.1 Individual Actions

Individuals can play a crucial role in supporting climate solutions through their actions and choices. Consider the following recommendations for making a positive impact:

- **Adopt Sustainable Practices**: Incorporate sustainable practices into your daily life, such as reducing energy consumption, minimizing waste, and supporting eco-friendly products. Small changes in your lifestyle can contribute to a larger collective impact.

- **Advocate for Change**: Use your voice to advocate for climate action at the local, national, and global levels. Engage in conversations about climate issues, support policy initiatives, and participate in community efforts to drive change.

- **Support Innovative Solutions**: Invest in or support innovative climate solutions, such as renewable energy projects, clean technology startups, and sustainable agriculture initiatives. Your support can help advance

these solutions and drive their
adoption.

4.3.2 Organizational Strategies

Organizations can contribute to climate
solutions by adopting sustainable
practices and supporting climate
initiatives. Consider the following
strategies for advancing organizational
climate action:

- **Implement Sustainable
  Practices**: Incorporate
  sustainability into your
  organization's operations,
  including energy efficiency
  measures, waste reduction, and
  sustainable sourcing. Develop
  policies and practices that align
  with environmental goals.
- **Foster Innovation**: Encourage
  innovation within your
  organization to develop new

solutions for addressing climate challenges. Support research and development efforts, and invest in technologies that contribute to sustainability.

- **Engage Stakeholders**: Collaborate with stakeholders, including employees, customers, and partners, to promote climate action and sustainability. Engage in dialogue about environmental goals and work together to achieve shared objectives.

4.3.3 Policy Recommendations

Policymakers play a critical role in shaping climate action through the development and implementation of policies. Consider the following recommendations for advancing climate policy:

- **Strengthen Climate Policies**: Advocate for robust climate policies that set ambitious targets for emissions reductions, promote renewable energy, and support climate resilience. Ensure that policies are aligned with scientific recommendations and address key climate challenges.
- **Support Innovation and Research**: Provide funding and support for research and innovation in climate science, technology, and solutions. Invest in projects that advance our understanding of climate change and develop new approaches for addressing it.
- **Promote Equity and Justice**: Ensure that climate policies address social inequalities and promote equitable outcomes.

Consider the needs of marginalized communities and support initiatives that enhance social justice and environmental equity.

4.4 Exercises and Reflections

Engaging in exercises and reflections can help deepen your understanding of the insights from visionary voices and inspire action. Consider the following activities to explore these perspectives further:

4.4.1 Activities to Explore Visionary Insights

- **Interviews with Thought Leaders**: Conduct interviews with thought leaders or experts in climate action. Gather their perspectives on key issues, challenges, and solutions. Use these insights to inform your understanding of climate solutions and strategies.

- **Analysis of Innovative Projects**: Research and analyze innovative climate projects and solutions. Evaluate their impact, effectiveness, and potential for replication. Consider how similar approaches could be applied in your own context.

- **Collaborative Discussions**: Participate in discussions or workshops with others interested in climate action. Share insights, exchange ideas, and collaborate on developing strategies for addressing climate challenges.

4.4.2 Reflection Questions

- Which visionary voices resonate with you the most, and why? How have their perspectives influenced your understanding of climate action?

- What are some examples of innovative projects or solutions that inspire you? How can you support or contribute to similar initiatives?
- How can you incorporate the insights from visionary voices into your own climate efforts? What practical steps can you take to apply these insights in your personal or professional life?
- What role do you think creativity and innovation play in addressing climate challenges? How can you foster a culture of innovation within your own community or organization?

Chapter 5: Practical Steps for a Better Future

5.1 Actionable Steps for Individuals

Individuals play a critical role in addressing climate change and promoting sustainability. By making conscious choices and taking proactive steps, individuals can contribute significantly to a better future. This section outlines practical actions that individuals can take to reduce their environmental impact and support climate solutions.

5.1.1 Reducing Personal Carbon Footprint

One of the most direct ways individuals can contribute to climate action is by reducing their carbon footprint. Here are some strategies for minimizing emissions:

- **Energy Efficiency**: Improve energy efficiency in your home by using energy-saving appliances, insulating your home, and adopting smart thermostats. Simple changes, such as switching to LED light bulbs and unplugging unused electronics, can reduce energy consumption.

- **Sustainable Transportation**: Reduce emissions by using public transportation, carpooling, biking, or walking instead of driving alone. Consider switching to an electric or hybrid vehicle to further decrease your carbon footprint.
- **Dietary Choices**: Adopt a more sustainable diet by reducing meat consumption, particularly red meat, and increasing the intake of plant-based foods. Support local and organic food producers to minimize the environmental impact of your food choices.

- **Waste Reduction**: Practice waste reduction by recycling, composting, and minimizing single-use plastics. Choose products with minimal packaging and participate in community cleanup efforts to reduce litter and waste.

5.1.2 Supporting Sustainable Products and Practices

Supporting businesses and products that prioritize sustainability can drive positive environmental change. Consider the following actions:

- **Eco-Friendly Products**: Choose products that are certified as eco-friendly, such as those with **Energy Star** or **Fair Trade** certifications. Look for products made from sustainable materials and produced using environmentally responsible practices.

- **Sustainable Fashion**: Support sustainable fashion brands that use ethical production methods and materials. Consider buying second-hand clothing and practicing mindful consumption to reduce the environmental impact of your wardrobe.
- **Green Investments**: Invest in companies and funds that prioritize environmental sustainability and social responsibility. Explore opportunities to support green technologies, renewable energy projects, and sustainable enterprises.
- **Community Engagement**: Get involved in local environmental initiatives and support

organizations working on climate and sustainability issues. Participate in community events, advocacy campaigns, and volunteer opportunities to contribute to local efforts.

5.1.3 Educating and Advocating for Change

Education and advocacy are essential for raising awareness and driving climate action. Here are some ways to get involved:

- **Raise Awareness**: Share information about climate change and sustainability with friends, family, and social networks. Use your platform to promote environmental issues and encourage others to take action.
- **Advocate for Policy Change**: Support and advocate for climate policies at the local, national, and

global levels. Engage with policymakers, participate in public consultations, and support campaigns that call for stronger environmental regulations.

- **Educate Others**: Offer workshops, seminars, or online content to educate others about climate solutions and sustainable practices. Share resources, tools, and strategies for individuals to implement in their daily lives.

5.2 Community-Based Actions and Initiatives

Communities play a crucial role in driving collective climate action and fostering sustainability. Collaborative efforts can amplify individual actions and create a more significant impact. This section explores community-based actions and initiatives that contribute to a better future.

5.2.1 Local Environmental Programs

Many communities have local programs and initiatives focused on environmental sustainability. Get involved in or support these programs to contribute to local climate efforts:

- **Community Gardens**: Participate in or support community garden projects that promote urban agriculture, local food production, and green spaces. Community gardens can provide fresh produce, foster community engagement, and enhance local resilience.
- **Recycling Programs**: Support and participate in local recycling programs to reduce waste and promote responsible disposal of materials. Advocate for improved

recycling facilities and programs in your community.

- **Energy Efficiency Initiatives**: Join or support initiatives aimed at improving energy efficiency in community buildings and public spaces. These initiatives may include retrofitting buildings, promoting energy-saving technologies, and raising awareness about energy conservation.

- **Tree Planting Campaigns**: Participate in tree planting campaigns to enhance urban greenery, improve air quality, and support local biodiversity. Tree planting events often involve community volunteers and contribute to environmental restoration efforts.

5.2.2 Collaborative Climate Projects

Collaborative climate projects bring together community members, organizations, and businesses to address environmental challenges. Consider participating in or supporting the following types of projects:

- **Climate Action Networks**: Join or support climate action networks that facilitate collaboration among stakeholders to address climate issues. These networks often focus on sharing resources, coordinating efforts, and driving collective action.
- **Sustainable Development Projects**: Support sustainable development projects that aim to improve infrastructure, promote economic growth, and enhance environmental sustainability. Projects may include renewable

energy installations, green building initiatives, and community resilience programs.

- **Citizen Science Initiatives**: Participate in citizen science projects that involve community members in collecting and analyzing environmental data. These projects can contribute to scientific research, increase public awareness, and inform policy decisions.
- **Local Environmental Advocacy Groups**: Engage with local advocacy groups that work on environmental issues, such as climate change, conservation, and social justice. These groups often organize campaigns, events, and outreach activities to promote environmental protection.

5.2.3 Building Resilient Communities

Building resilient communities involves strengthening the ability of communities to adapt to and recover from environmental and climate-related challenges. Here are some strategies for fostering community resilience:

- **Disaster Preparedness**: Develop and participate in disaster preparedness programs that equip communities to respond to natural disasters and climate impacts. These programs may include emergency response plans, evacuation procedures, and community training.
- **Resilient Infrastructure**: Support efforts to build and maintain resilient infrastructure that can withstand climate-related hazards. This includes investing in flood protection, stormwater

management, and sustainable urban planning.

- **Social Support Networks**: Strengthen social support networks within communities to provide assistance during times of crisis. Building strong relationships and networks can enhance community cohesion and resilience.

5.3 Organizational Strategies for Sustainability

Organizations have a significant impact on the environment through their operations, products, and services. By adopting sustainable practices and supporting climate initiatives, organizations can contribute to a better future. This section outlines strategies for organizations to advance sustainability and climate action.

5.3.1 Implementing Sustainable Business Practices

Organizations can reduce their environmental impact by implementing sustainable business practices across various aspects of their operations:

- **Energy Management**: Adopt energy management practices to reduce energy consumption and greenhouse gas emissions. This may include energy audits, efficiency upgrades, and the use of renewable energy sources.
- **Sustainable Supply Chains**: Develop sustainable supply chain practices by sourcing materials and products from responsible suppliers. Implement criteria for environmental and social responsibility in procurement decisions.
- **Waste Reduction and Recycling**: Implement waste reduction and

recycling programs to minimize waste generation and promote responsible disposal. Consider initiatives such as zero waste programs and circular economy practices.

- **Green Certifications**: Pursue green certifications, such as **LEED** (Leadership in Energy and Environmental Design) or **B Corp** certification, to demonstrate a commitment to sustainability and environmental stewardship.

5.3.2 Promoting Employee Engagement and Culture

Engaging employees and fostering a culture of sustainability can enhance organizational commitment to climate action. Consider the following strategies:

- **Sustainability Training**: Provide training and resources to employees on sustainability practices and climate action. Educate staff about the organization's sustainability goals and how they can contribute to achieving them.

- **Employee Involvement**: Encourage employee involvement in sustainability initiatives and decision-making. Create opportunities for staff to participate in green projects, volunteer activities, and sustainability committees.
- **Recognition and Rewards**: Recognize and reward employees for their contributions to sustainability efforts. Implement incentive programs that encourage environmentally responsible behaviors and innovations.
- **Transparent Communication**: Maintain transparent

communication about the organization's sustainability goals, progress, and challenges. Share updates with employees and stakeholders to foster trust and engagement.

5.3.3 Supporting Community and Global Initiatives

Organizations can extend their impact by supporting community and global initiatives focused on climate action and sustainability:

- **Partnerships and Collaborations**: Partner with local and global organizations to support climate initiatives and sustainable development projects. Collaborate on research, advocacy, and project implementation to amplify impact.
- **Corporate Social Responsibility (CSR)**: Integrate sustainability into corporate social responsibility

programs by supporting environmental and social causes. Invest in initiatives that align with the organization's values and contribute to positive environmental and social outcomes.

- **Global Climate Agreements**: Align organizational goals with global climate agreements and initiatives, such as the **Paris Agreement**. Support international efforts to address climate change and promote global sustainability.

5.4 Exercises and Reflections

Engaging in exercises and reflections can help individuals, communities, and organizations apply practical steps for a better future. Consider the following activities to explore these strategies further:

5.4.1 Activities to Implement Practical Steps

- **Personal Action Plan**: Develop a personal action plan outlining specific steps you can take to reduce your environmental impact and support climate solutions. Set measurable goals and track your progress over time.

- **Community Project Planning**: Plan and implement a community-based project focused on sustainability or climate action. Collaborate with local stakeholders, identify project objectives, and develop a plan for execution.

- **Organizational Sustainability Assessment**: Conduct an assessment of your organization's sustainability practices and identify areas for improvement. Develop an action plan to enhance sustainability and engage employees in the process.

- **Advocacy Campaign**: Design and execute an advocacy campaign to raise awareness about climate issues and promote policy change. Use various communication channels to reach your target audience and encourage action.

5.4.2 Reflection Questions

- What are some specific actions you can take to reduce your personal carbon footprint? How can you incorporate these actions into your daily routine?
- How can you support and engage with local environmental initiatives and organizations? What opportunities are available in your community?

- What strategies can your organization implement to advance sustainability and climate action? How can you involve employees and stakeholders in these efforts?
- How can you contribute to global climate initiatives and support international efforts to address climate change? What role can you play in advancing global sustainability goals?

Chapter 6: Reflecting on Possibilities and Actions

6.1 The Importance of Reflection in Climate Action

Reflection is a crucial component of effective climate action and sustainability efforts. By taking the time to evaluate our actions, understand their impacts, and adjust our strategies, we can ensure that our efforts are both meaningful and effective. This section explores the role of reflection in climate action and provides practical guidance on how to incorporate it into your sustainability efforts.

6.1.1 Understanding the Impact of Your Actions

Reflecting on the impact of your actions helps to evaluate their effectiveness and identify areas for improvement. Consider the following aspects:

- **Quantitative Assessment**: Measure the tangible outcomes of your actions, such as reductions in energy consumption, waste generation, or carbon emissions. Use data and metrics to assess progress and identify successes and challenges.
- **Qualitative Evaluation**: Evaluate the qualitative aspects of your actions, such as changes in behavior, attitudes, and community engagement. Assess the broader impact on social, cultural, and environmental aspects.
- **Feedback and Learning**: Seek feedback from others, including peers, experts, and community members. Use this feedback to learn from your experiences and

make informed decisions about future actions.

6.1.2 Setting Goals and Measuring Progress

Setting clear, measurable goals and tracking progress is essential for effective climate action. Here are some strategies for goal-setting and progress monitoring:

- **SMART Goals**: Set **SMART** (Specific, Measurable, Achievable, Relevant, Time-bound) goals to guide your climate action efforts. Ensure that your goals are clearly defined and include criteria for measuring success.
- **Tracking Tools**: Use tracking tools, such as journals, apps, or spreadsheets, to monitor your progress toward your goals. Regularly update and review your tracking tools to stay informed

about your achievements and challenges.

- **Periodic Reviews**: Conduct periodic reviews of your progress to assess whether you are meeting your goals and to identify areas for adjustment. Use these reviews to reflect on your strategies and make necessary changes.

6.1.3 Adjusting Strategies Based on Reflection

Reflection often leads to insights that can inform adjustments to your strategies. Consider the following approaches:

- **Reevaluating Priorities**: Based on your reflections, reevaluate your priorities and focus areas. Adjust your goals and strategies to align with new insights or changing circumstances.

- **Adapting Actions**: Modify your actions and approaches to address any identified gaps or challenges. Implement new strategies or tactics to enhance your effectiveness and impact.
- **Continuous Improvement**: Embrace a mindset of continuous improvement by regularly reflecting on your efforts and seeking opportunities for growth and development. Stay open to new ideas and approaches that can enhance your climate action efforts.

6.2 Engaging with Others to Enhance Impact

Engaging with others, including community members, organizations, and stakeholders, can enhance the impact of your climate action efforts. Collaborative engagement fosters shared learning, strengthens partnerships, and amplifies collective impact. This section explores ways to engage with others and build supportive networks.

6.2.1 Building Collaborative Partnerships

Collaborative partnerships can enhance the effectiveness and reach of your climate action efforts. Consider the following strategies for building and maintaining partnerships:

- **Identifying Partners**: Identify potential partners, including individuals, organizations, and businesses, that share similar goals and values. Look for opportunities to collaborate on projects, initiatives, and advocacy efforts.
- **Establishing Agreements**: Develop formal or informal agreements with partners to outline roles,

responsibilities, and expectations. Ensure that agreements are clear and mutually beneficial.

- **Fostering Communication**: Maintain open and transparent communication with your partners. Regularly share updates, provide feedback, and discuss any challenges or opportunities for collaboration.

6.2.2 Participating in Collaborative Initiatives

Participating in collaborative initiatives can expand your impact and contribute to collective climate action. Explore the following opportunities:

- **Joint Projects**: Join or initiate joint projects that address climate challenges and promote sustainability. Collaborate with partners to design, implement, and evaluate projects that align with shared goals.

- **Networking Events**: Attend networking events, conferences, and workshops to connect with others in the climate action and sustainability community. Use these opportunities to learn from others, share your experiences, and build relationships.
- **Community Engagement**: Engage with your local community to promote climate action and sustainability. Participate in community events, outreach activities, and educational programs to raise awareness and foster collective action.

6.2.3 Leveraging Collective Resources

Leveraging collective resources can enhance your ability to address climate challenges and achieve your goals. Consider the following approaches:

- **Resource Sharing**: Share resources, such as tools, knowledge, and expertise, with your partners and community members. Collaborative resource sharing can increase efficiency and effectiveness.
- **Funding and Support**: Seek funding and support from partners, grant programs, and philanthropic organizations to support your climate action efforts. Explore opportunities for financial and in-kind contributions.
- **Knowledge Exchange**: Participate in knowledge exchange activities to share and gain insights on climate solutions and best practices. Engage in discussions,

workshops, and training sessions to expand your understanding and capabilities.

6.3 Personal and Professional Growth Through Reflection

Reflection can also contribute to personal and professional growth. By understanding your experiences and learning from them, you can enhance your skills, knowledge, and impact. This section explores ways to leverage reflection for personal and professional development.

6.3.1 Learning from Experiences

Reflect on your experiences to gain valuable insights and lessons. Consider the following approaches:

- **Reflective Practice**: Engage in reflective practice by regularly reviewing your experiences and identifying key learnings. Use reflective journaling or discussions with mentors to deepen your understanding.

- **Identifying Strengths and Areas for Growth**: Assess your strengths and areas for growth based on your reflections. Use this knowledge to build on your strengths and address areas where improvement is needed.
- **Applying Lessons Learned**: Apply the lessons learned from your reflections to future actions and decisions. Use these insights to enhance your approach and increase your effectiveness.

6.3.2 Developing New Skills

Reflection can inform your development of new skills and competencies. Explore the following strategies:

- **Skill Assessment**: Assess your current skills and identify areas where additional development is needed. Seek opportunities for training, education, and skill-building in areas related to climate action and sustainability.
- **Professional Development**: Pursue professional development opportunities, such as workshops, certifications, and courses, to enhance your expertise and knowledge. Engage in continuous learning to stay informed about emerging trends and practices.
- **Mentorship and Coaching**: Seek mentorship and coaching from experienced professionals in the climate action and sustainability field. Learn from their experiences and insights to support your growth and development.

6.4 Exercises and Reflections

Engaging in exercises and reflections can help you apply the concepts from this chapter and deepen your understanding of climate action and sustainability. Consider the following activities:

6.4.1 Reflection Exercises

- **Personal Reflection Journal**: Maintain a personal reflection journal to document your experiences, insights, and progress related to climate action. Use the journal to review your actions, assess their impact, and plan for future steps.
- **Goal Review Session**: Conduct a goal review session to evaluate your progress toward your climate action goals. Reflect on your achievements, challenges, and

lessons learned, and adjust your goals as needed.

- **Collaborative Reflection**: Engage in collaborative reflection with partners or community members to share experiences and insights. Discuss successes, challenges, and opportunities for improvement to enhance collective impact.

6.4.2 Action Planning

- **Action Plan Development**: Develop an action plan based on your reflections and insights. Outline specific steps, goals, and timelines for implementing changes and addressing identified challenges.
- **Skill Development Plan**: Create a skill development plan to identify and pursue opportunities for building new skills and

competencies related to climate action and sustainability.

- **Community Engagement Plan**: Design a community engagement plan to enhance your involvement in local climate initiatives and collaborative projects. Outline strategies for building partnerships and supporting community efforts.

_____

_____

_____

_____

_____

_____

_____

Made in the USA
Las Vegas, NV
24 October 2024